SAE EDGE™ RESEARCH REPORT

Unsettled Topics Concerning Autonomous Public Transportation Systems

Rahul Razdan, PhD
RazTech Research Institute

EDGE DEVELOPMENT TEAM

Mustafa Akbas, PhD, *Embry-Riddle Aeronautical University*

Jim Cherian, PhD, *Centre of Excellence for Testing & Research of Autonomous Vehicles NTU*

Niels de Boer, *Centre of Excellence for Testing & Research of Autonomous Vehicles NTU*

Bernard Schmidt, *Jacksonville Transportation Authority*

Ravio Sell, PhD, *Tallinn University of Technology*

Clayton Tino, PhD, *Beep, Inc.*

SAE INTERNATIONAL®

Warrendale, Pennsylvania, USA

About the Publisher

SAE EDGE™ Research Report Disclaimer

SAE EDGE™ Research Reports focus on topics that are dynamic, in which knowledge is incomplete, and which have yet to be standardized. They represent the collective wisdom of a group of experts and serve as a practical guide to the reader in understanding unsettled subject matter. They are not meant to provide a recommended practice or protocol. The experts have assembled as a community of practitioners to contribute and collectivize their thoughts and points of view; these are not the positions of the institutions or businesses with which they are affiliated, nor is one contributor's perspective advanced over other contributors. SAE EDGE™ Research Reports are the property of SAE International and SAE alone is responsible for their content.

About This Publication

SAE EDGE™ Research Reports provide state-of-the-art and state-of-industry examinations of the most significant topics in mobility engineering. SAE EDGE™ contributors are experts from research, academia, and industry who have come together to explore and define the most critical advancements, challenges, and future direction in areas such as vehicle automation, unmanned aircraft, cybersecurity, advanced propulsion, advanced manufacturing, Internet of Things, and connectivity.

Related Resources

SAE MOBILUS® Automated & Connected Knowledge Hub
https://saemobilus.sae.org/automated-connected/

SAE Team

Frank Menchaca, Chief Growth Officer
Michael Thompson, Director of Standards, Information and Research Publications
Monica Nogueira, Director of Content Acquisition
Beth Ellen Dibeler, Product Manager
William Kucinski, Managing Technical Editor

Printed in USA

EPR2020020
ISSN 2640-3536
e-ISSN 2640-3544
ISBN 978-1-4686-0248-7

To purchase bulk quantities, please contact: SAE Customer Service

E-mail: CustomerService@sae.org
Phone: 877-606-7323 (*inside USA and Canada*)
+1-724-776-4970 (*outside USA*)
Fax: +1-724-776-0790

https://www.sae.org/publications/edge-research-reports

About the Editor

Dr. Rahul Razdan is a seasoned scientist and business executive who has had significant roles in the world of academia, startups, and Fortune 500 companies. In corporate roles, he was the Senior Vice President of Strategy at Flextronics, General Manager of the System and Functional Verification Business at Cadence Design Systems, and Alpha CPU architect at Digital Equipment Corporation. He has built successful startups in areas such wireless power, machine learning, and low-power electronics design. In academia, beyond a large range of worldwide collaborations, he maintains academic appointments at the University of Florida, Florida Polytechnic University, and Institute for Human Machine Cognition.

Today, he is CEO of Raztech Research Institute (www.razinstitute.com), a technology-based research institute, which incubates ideas at the intersection of technology and society. Active projects for the institute include avvc.net in the area of autonomous vehicles, nextgenedu.net in the area of STEM education, and Anew-da.com in the area of EDA for long life cycle products. He has a bachelor's (1984) and master's (1985) of science from Carnegie Mellon University and earned his doctorate at Harvard University (1994).

contents

Unsettled Topics Concerning Autonomous Public Transportation Systems

Abstract

With the launch of the Defense Advanced Research Projects Agency (DARPA) Grand Challenges in 2004 and 2005, autonomous driving technology has been slowly making its way toward commercialization. Billions of dollars have been invested in the development of the core technology and potential use models. While the technology required for a fully functional automated vehicle (AV) is quite a few years away from reality, specific restricted-use models such as robo-taxis, truck convoying, and advanced driver assistance systems (ADAS) show great promise. Another use model, which also shows a great deal of potential, is the use of AV technology for public transportation.

Technologically, *public transportation shuttles are an ideal potential application as they operate at low speeds on fixed routes*. From a business point of view, there is already an existing public transit business model, and there are potential opportunities for providing superior quality of service for disadvantaged communities. However, most of the utility is typically found in dense urban environments, which can present unique challenges. Further, *the operators of public transportation perform within a public sector regime, which has additional requirements*. Thus, it is quite instructive to look at the divergent experiences of leading cities as they engage with AV technology for their specific purposes. In this SAE EDGE Research Report, we will examine the still unsettled topics and generic solutions for the public transportation market and the unique AV deployment schemas for the cities of Jacksonville, Florida; Singapore; Tallinn, Estonia; and finally Orlando/Lake Nona, Florida.

NOTE: SAE EDGE™ Research Reports are intended to identify and illuminate key issues in emerging, but still unsettled, technologies of interest to the mobility industry. The goal of SAE EDGE™ Research Reports is to stimulate discussion and work in the hope of promoting and speeding resolution of identified issues. SAE EDGE™ Research Reports are not intended to resolve the challenges they identify or close any topic to further scrutiny.

RAHUL RAZDAN, PhD
RazTech Research Institute

EDGE Development Team

Mustafa Akbas, PhD, *Embry-Riddle Aeronautical University*
Jim Cherian, PhD, *Centre of Excellence for Testing & Research of Autonomous Vehicles NTU*
Niels de Boer, *Centre of Excellence for Testing & Research of Autonomous Vehicles NTU*
Bernard Schmidt, *Jacksonville Transportation Authority*
Ravio Sell, PhD, *Tallinn University of Technology*
Clayton Tino, PhD, *Beep, Inc.*

ISSN 2640-3536

Introduction

Public transportation systems are vital to the mobility needs for cities and towns. Within the city core, it is not unusual to find four to six vastly different modes of transportation operating concurrently. Table 1 shows the variety of these modes, their capacity to move people, the variation in speed, and the road surface consumed. While in transit, these modes may be somewhat isolated from each other, but they often interact at points of transfer for passengers to embark/disembark. To facilitate this process, specialized transfer zones are often established to optimize traffic and commuter safety. Any public transportation autonomy solution must account for the existing environment and integrate smoothly into this very organic ecosystem.

In the business of transportation, the dominant form factors in terms of volume and investment are the traditional passenger cars built by popular automotive organizations. These original equipment manufacturers (OEMs) build their capabilities for design, manufacturing, and distribution for the realities of the passenger market. In contrast, public transport solutions such as passenger buses, streetcars, or subways are specialized (often customized solutions) and have entirely different characteristics and are served by a different set of players.

With the evolution of the AV market, a similar differentiation has occurred. While traditional automotive OEMs, Tesla, and even Waymo focused on the traditional passenger automobile marketplace, a newer set of corporations have emerged to service the public transportation marketplace with a somewhat standard form factor (Figure 1).

Table 2 lists a set of current vendors offering solutions in the public transportation space. They consist of specialized solution providers, such as Easymile, Navya, and Proterra. Traditional Tier 1 players such as Bosch and Continental have also built demo vehicles—and there is even participation from OEMs such as Daimler [1].

Public transportation agencies are familiar with working with specialized suppliers for their equipment, but designing a solution with autonomous capability has its challenges [2, 3]:

1. Road Architecture: Given the complexity of urban cores, what is the right architecture for introducing AVs?
2. Public Value: What is the real additional public value for introducing AV technology?
3. Operational Integration: How do other operational elements, such as emergency services, maintenance, and customer satisfaction integrate into the solution?
4. Supply Chain: Given the nascent nature of the marketplace and the relative financial fragility of the leading technology providers in the AV shuttle space, how does one manage supply chain risk?
5. Safety: How does one manage safety at all levels (e.g., mechanical (crash worthiness), perception (effects of weather), or decision making (logical)?

In this report, we will examine the details of four cities' engagement (Jacksonville, Florida; Singapore, Tallinn, Estonia; and Orlando/Lake Nona, Florida) with autonomous technology, and glean insights on the broader lessons and challenges that the AV industry is facing:

1. With Jacksonville, our journey is led by Bernard Schmidt and the Automation team at the Jacksonville Transportation Authority. Bernard is an Amazon veteran who is developing the Ultimate Urban Connector (UUC) project.
2. With Singapore, our journey is led by Dr. Jim Cherian and Mr. Niels de Boer. Both work at the Centre of Excellence at Nanyang Technological University. In these roles, they are intimately connected with Singapore's AV efforts.
3. With Tallinn, our journey is led by Dr. Ravio Sell. Sell headed a research group that built an open-source-based AV shuttle in about a year. This shuttle is in the process of being commercialized and is currently deployed for testing with exisiting public transportation solutions in Tallinn.
4. With Orlando/Lake Nona, our journey is led by Clayton Tino, the Chief Technology Officer at Beep. Beep is a unique AV service operator, that has rolled out a functional AV service in the Lake Nona planned community.

TABLE 1. Capacity, speed, road width of transportation modes.

Type of movement	Capacity (per hour)	Average speed (mph)	Road width (ft)
Private Auto	600-1600	20-35	10-15
Protected Bike Lane	4000	8-15	4-5
Dedicated Transit	4000-8000	8-15	10-15
Pedestrian Sidewalk	8000-9000	2-4	4-5
On-Street transit way	10,000-25,000	15-20	10-15

Data taken from Nacto.org. RazTech Research Institute.

FIGURE 1. **Typical form factor for an AV shuttle.**

SariMe/Shutterstock.com

Jacksonville and the Ultimate Urban Circulator Project

Jacksonville is one of the most populous cities in Florida and is the largest city by land mass in the continental US. It boasts several major military bases as well as one of the largest civilian deepwater ports in the world. Much like the rest of the state of Florida, Jacksonville has been growing rapidly and this has driven development as well as the associated needs for transportation services (Figure 2).

The Jacksonville Transportation Authority (JTA) is a mid-size transit agency that began as an expressway authority, that now provides mobility services. The JTA has operated an automated elevated people mover called the Skyway Express (Skyway) since 1989. At 2.5 miles, it serves the core downtown district and it connects the north and south banks of Jacksonville across the St. Johns River. With the city's population projected to continue to increase, the public transportation challenge now includes connecting newly developed neighborhood centers, such as the entertainment and hospital districts. With the aging infrastructure and obsolescence of the existing Skyway, the JTA determined that it was the perfect time to consider AV technology as a compelling upgrade to the system.

TABLE 2. **AV shuttle makers.**

AB Volvo	Bosch	Coast Autonomous	Continental
Denso	Daimler	Baidu	Local Motors
Proterra	Easymile	Navya	Scania

RazTech Research Institute

AV technology has the characteristics to provide a viable and cost-effective solution for converting a system like the Skyway network without substantial infrastructure changes. Through a task force assembled in 2015 with various JTA, City of Jacksonville, business, and community stakeholders, an idea took hold that AVs could reuse the old Skyway's civil infrastructure and then transition onto existing surface streets to extend the network (Figure 3). This vision led to the Ultimate Urban Circulator (U^2C) [4] project.

While promising for Jacksonville, AVs are still an emerging technology with many open questions:

- Can AV shuttles handle mixed traffic?
- Will they require dedicated lanes and, if so, will there be additional infrastructure investments?
- What will be the ongoing maintenance requirements?
- Will the vehicles still need drivers?
- Who will provide the answers to these questions?

FIGURE 2. **Jacksonville, Skyline of Jacksonville, Florida.**

ESB Professional/Shutterstock.com

FIGURE 3. **First Phase Bay Street Innovation Corridor AV Route.**

Reprinted with permission from Jacksonville Transportation Authority

For public transportation entities, the traditional operating model is to engage with a large network of civil engineering firms through the procurement cycle. But would transit staff have the expertise to put together a Request for Information and, if so, would traditional civil engineering firms even have the capability to credibly offer a solution in response? Interestingly, one of the early key decisions was to hire experts from Amazon who could formulate the right questions [5].

Armed with a team, which was familiar with new technology introduction, the automation group of the JTA started by building a facility (an isolated road in downtown Jacksonville) to test vendor solutions. This was wise, because they soon discovered there was a big difference between the demo-like capabilities presented by the AV vendors and the robust solution required by a public transportation utility. To resolve these issues, the automation division works with vendors in an unusual research and development-level collaboration. Understanding the criticality of the test-and-learn program and vendor relationships, a local governance structure was established and dubbed as the Bay Street Innovation Council (BSIC). The BSIC was initiated to bring key government stakeholders to the table to reach alignment on the first phase of the U²C program and overall execution.

After procuring a relatively robust AV, the next task was to understand all the interesting use models:

- Should there be a driver?
- What about access for the disabled?
- What about maintenance and energy?

Since there were no precedents, each of these topics had to be investigated and solutions were developed.

One major area of concern was ensuring that the JTA provided excellent customer service and a safe environment for riders. An important challenge was the topic of safety and certification. Another was integration with first responders. In 2019, the JTA hosted the first Northeast Regional First Responders Training for Autonomous Vehicles, which set the groundwork for the JTA to establish a First Responder Council. This new body brought local law enforcement and the first responder community together to help refine the U²C strategy moving forward. This work was completed under the legal framework created by the state of Florida's forward-looking AV bills (HB 311) [6], which formed the framework for AV manufacturers to test autonomous technology within the state. The JTA has done considerable work in advancing the Bay Street Innovation Project (the first phase of the U²C). As a result of this advancement, the project is an active two-step procurement process, which will result in contract award in early 2021. The Bay Street Innovation Corridor project utilizes the JTA's Better Utilizing Investments to Leverage Development (BUILD) Transportation Discretionary Grant program funding from the US Department of Transportation.

Recommendations

1. **AV Education:** The existing transportation ecosystem consisting of civil engineering consulting firms and traditional transportation agencies have developed a well-honed structure for finishing elaborate road projects. However, the newer electronic concepts such as advanced sensors, radio frequency communications, and machine learning are not well understood. The transportation ecosystem would be well served to invest in the education of its staff on these issues.
2. **Engagement Model:** In the traditional public transportation realm, agencies bid out contracts to vendors and then hold vendors to performance targets. This process works reasonably well for situations where the technology is well established. However, AV technology is in a developmental stage, and public agencies engaging in projects in this area need to have a mindset much more attuned to advanced product development. This means investment in on-boarding and testing.

Singapore and Integrated Transportation Planning

Singapore is a dense city-state with a large growing population (slightly under 6,000,000). In 2015, the Singapore government launched the Sustainable Singapore Blueprint (SSB) [7], which had five strategies to build a sustainable city. One of the key elements in the SSB is "a car-lite" Singapore. Thus, SSB focuses on shifting mobility to walking, biking, and public transportation (Figure 4).

Singapore's Case for AV Technology

Given the increasingly point-to-point nature of travel for both passengers and logistics, there is a tradeoff between large buses (perhaps on inefficient fixed routes) and smaller shuttles with a more dynamic responsive transportation system. This need for point-to-point transportation is amplified for a rapidly aging population. In fact, the challenge for Singapore is that both public transport and logistics (internet shopping and related home deliveries are increasing) are highly constrained by the availability of drivers.

In 2018, the Deputy Secretary of Land and Corporate at the Ministry of Transport, Lee Chuan Teck, published a far-reaching and visionary article "Public Transport Planning and the Technological Revolution" [8], which includes the following critical elements:

- Freeing public transportation from fixed routes and schedules
- Using AVs to overcome manpower constraints
- Narrowing the gap between private and public transportation

FIGURE 4. **Singapore skyline.**

Sean Pavone/Shutterstock.com

Most importantly, there was a recognition that new technologies create new challenges. In the article, Deputy Secretary Teck mentioned that AVs would require financial support and regulatory sandboxes to facilitate experimentation. Deputy Secretary Teck also recognized the potential shift in labor and suggested that drivers may instead be trained to handle exceptions, such as when the autonomous buses break down or commuter incidents happen.

Turning AV Vision into a Plan, which Ensures Safety and Builds a Framework for Regulation

Similar to Jacksonville [9], a joint team was developed between the Centre of Excellence for Testing & Research of Autonomous Vehicles at Nanyang Technological University (CETRAN), the Land Transport Authority, traffic police, and JTC Corporation (planning agency). CETRAN [10, 11]—one of the world's top-rated universities—thus, a critical linkage was forged between operators, regulators, and technology experts. The Land Transport Authority falls under the Ministry of Transport, and the overall initiative falls under the Smart Nation Program, which is run directly from the Prime Minister's office. Given the broader potential impacts of AVs beyond transportation, the Committee for Autonomous Road Transport Singapore (CARTS) [12] was created to include representatives from other ministries, agencies, and also from industry.

Quite sensibly, the objective of the framework was to support trials with a testing regime, which links an increase in the test area to demonstrated capabilities. Any vehicle currently running on public roads has gone through the CETRAN AV Test Centre (which includes a custom-built test track [13] that also encompasses urban public transport-relevant features, such as bus stops with and without bus bays) to assess its readiness in terms of safety for such trials. Currently, all testing requires an onboard safety driver. This structure gives assurance to the public and provides a structure for the regulator to learn about the technology and build effective long-term regulations. As the Ike report [14] points out, artificial intelligence (AI) machines can have robustness issues and an inclination to "learn only the test." Thus, the details of the framework tests are not publicly available. In 2019, the area available for the testing framework was expanded to Western Singapore [15].

In terms of capital, the Singapore government has awarded funding for AV trials to a number of organizations such as Motional (Aptiv, Delphi, NuTonomy), Singapore Technologies, and NTU-Volvo. In January, they opened a call for collaboration [16], which includes a progressive development approach consisting of research and development, trial, and pilot program phases.

To date, CETRAN has worked with 10 different AV developers and more than 30 vehicles have been tested for the purpose of acquiring public AV trial authorization. While the largest fleet is from Motional, there are also developers operating single vehicles, ranging from small cars based on the Renault Zoe and Mitsubishi i-MiEV to full-size city buses from NTU-Volvo and ST Engineering Land Systems.

Singapore's current public transport routes already admit passengers at least in a limited manner, while conducting trials (still with a human safety driver). Two well-known examples are the Sentosa bus trials [17, 18] and Comfort Delgro's Easymile EZ10 trials at National University of Singapore's [19, 20] "NUSmart Shuttle service" campus (Figure 5).

Recommendations

1. **Master Planning:** As a city-state, Singapore has the ability to integrate regulation, agencies, research institutions, and even funding sources. With that backdrop, Singapore was equipped to develop a plan highly specific to their needs. A clear utility of such a process is clarity regarding the exact expected benefits of technology introduction and anticipation of handling disruptions smoothly.
2. **Staged Engagement Model:** Similar to Jacksonville, Singapore focused on a staged engagement process to understand the technology, use model, and implications for regulation. Defining a stage-gate process is definitely recommended.
3. **Integrated Team Approach:** Building a working team, which integrates the functional elements with the deep involvement of a technical university is critical to building a stage-gate process and understanding risks.

Tallinn and Leveraging Open Source

In technology, open-source environments have been one of the most important organizational models in the last 30 years. Perhaps the most successful of these has been the Linux operating system, borne into a world of proprietary operating systems. These proprietary operating systems provided the interface between hardware developers and application software developers; however, their proprietary nature inhibited hardware and application engineering progress. In 1991, Linux was introduced as an open-source platform and grew in popularity. Today, Linux has significant market share in the commercial landscape with an ecosystem larger than anything any single company—or even country—can build. Linux's seminal moment occurred in 2000 when IBM announced its full support of Linux—effectively deemphasizing the path of their own proprietary operating systems. Android, WordPress, Apache, and countless other open-source environments show the viability of this organizational structure [21].

FIGURE 5. **Two AV trial routes toward automated public transport in Singapore: (a) Sentosa bus trials [18] and (b) NUSmart Shuttle service [20].**

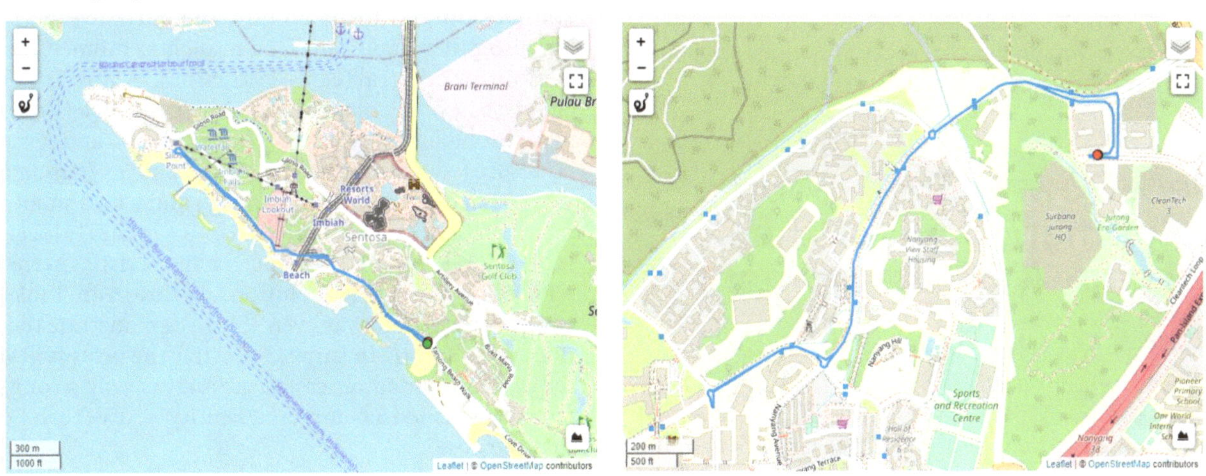

(a) (b)

FIGURE 6. Tallinn University of Technology campus operations.

Reprinted with permission from Taltech

Building a Functioning AV Shuttle with a Small Team Using Open-Source Methods

A research team from the Tallinn University of Technology (TalTech) [22] in Estonia took on the challenge of building an AV shuttle using open-source solutions. The result was the ISEAUTO [23] shuttle (Figure 6).

TalTech is a leading polytechnic university in Estonia. In the world of technology, Estonia is best known for providing some of the key founders of Skype. Like many AV design teams, TalTech's path started with robotics competitions such as Robotex [24]. In June 2017, an ambitious project was started with the goal to develop a low-speed AV shuttle for the university anniversary in September 2018. TalTech invested in the project because they believed developing AV technology was a multifaceted challenge that every ambitious academic community at the university could contribute toward.

Scaled-Down, Open-Source Development Amidst the Corporate AV Development Boom

The project was a joint effort between TalTech [25] and Silberauto Estonia [25], a distributor of automobiles. Silberauto focused on body design and mechanical assembly, while the TalTech team took on the daunting task of providing the entire AV electronics and software subsystem. The team consisted of approximately 18 students working with dedicated faculty. Overall, a relatively small team had to implement all the major AV functionality (Figure 7).

Due to time and cost constraints, the hardware components were off-the-shelf (Velodyne Lidar, standard PC with Nvidia graphics cards, Point Grey cameras, etc.). The key challenge was the software system, which had to be customized for the application and location. The only viable path was to start with a base consisting of open-source environments. However, which one?

An investigation [26] of open-source environments revealed four major candidates: Nvidia Drive [27], OpenPilot [28], Baidu Apollo [29], and Autoware [30]. While Autoware was sponsored by a research group from Nagoya University in Japan, the others were sponsored by commercial entities.

When choosing an open-source environment, the evaluation should not focus on only the current functionality, but also other important factors such as momentum of the community. In TalTech's evaluation, Autoware [31] was chosen based on completeness, independence (from particular automobiles or electronics hardware), and governance structure (company-led versus independent foundation).

Using the open-source strategy and with a small student team, a solution was built in time for the TalTech anniversary [32]. The performance characteristics were in line with other "last-mile" solutions as shown in Table 3 [33]. Uniquely, the team was even able to demonstrate the operation of the shuttle in snow [34]. After the prototype launch, the demand for commercialization was very strong, so the team established a commercial entity—AuVeTech [35]—for that purpose.

According to the team leader [36], the immediate takeaway in building an AV is that it is not impossible and

FIGURE 7. **AV high-level architecture.**

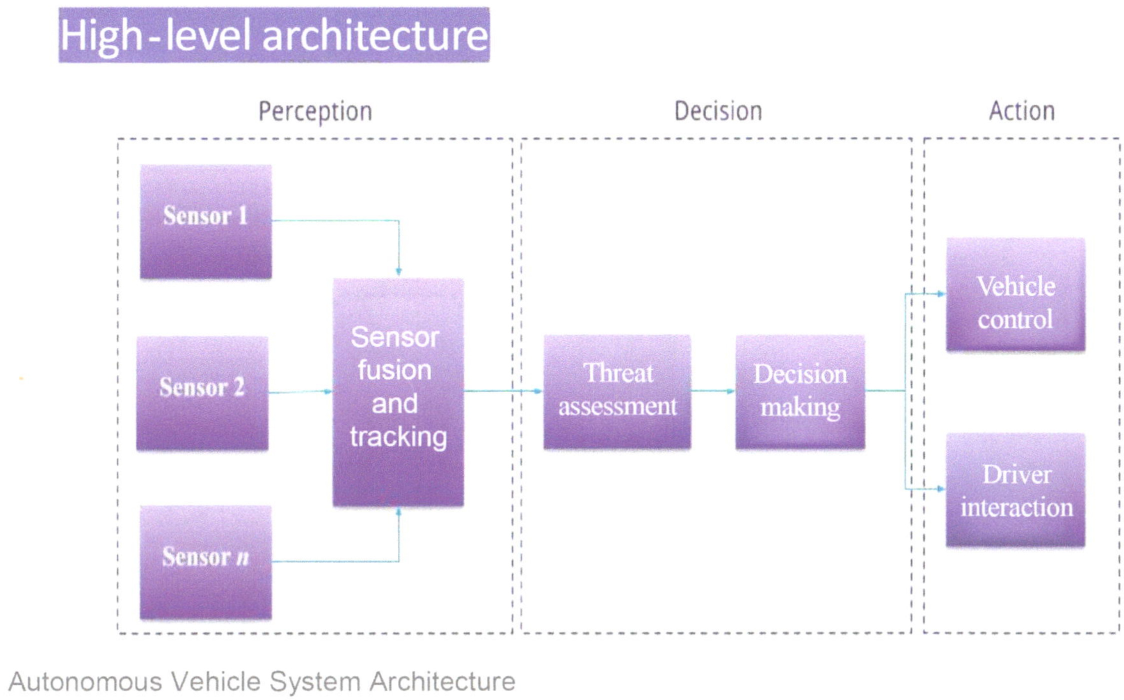

the task shouldn't be limited to only big OEMs. A key to the success was the availability of open-source software upon which the team could build their unique system design.

Building a research prototype with students is a reasonably common story. As Stefan Seltz-Axmacher, the CEO of Starsky Robotics, points out [37], building something that "works" is an accomplishment. However, building a true product that is statistically reliable requires ten to a thousand times more work.

Bending the Cost Curve with the Open-Source Approach

Recently, the TalTech team, with commercial partners, launched a test run at the Tallinn business campus in Ulemiste City (Figure 8), using three buses with smart bus stops (real-time location and access). The goal of the current tests is to collect real-life experience and continue developing the buses as a natural extension of public transportation. The route has some complexity with tram track crossings and complex interactions with other users. The overall work is funded by Future Automated Bus Urban Level Operation Systems (FABULOUS) from the European Union and is coordinated by Forum Virium Helsinki.

Recommendations

1. **Open-Source and Safety:** Open-source is a very powerful technology innovation engine, and it has been fruitfully used in traditional information

TABLE 3. **ISEAUTO shuttle results.**

Name (Developer)	Mspeed/Capacity/Range (km)	Software
Apollo minibus (Baidu/King Kong)	40/14/130	Baidu Apollo
EZ10 (EasyMile)	45/15/80	Custom
Olli (Local Motors)	40/12/60	IBM Watson AI
Parkshuttle (2GetThere)	40/22/100	Custom
Navya	45/15/xx	Custom
Novus drive (Hi-Tech Robotic Systemz)	40/xx/140	Custom
Wepod (TU Delft)	25/6/100	ROS Based
ISEAUTO	20/6/100	ROS Autoware

FIGURE 8. **Public shuttle route for Lake Nona, Florida.**

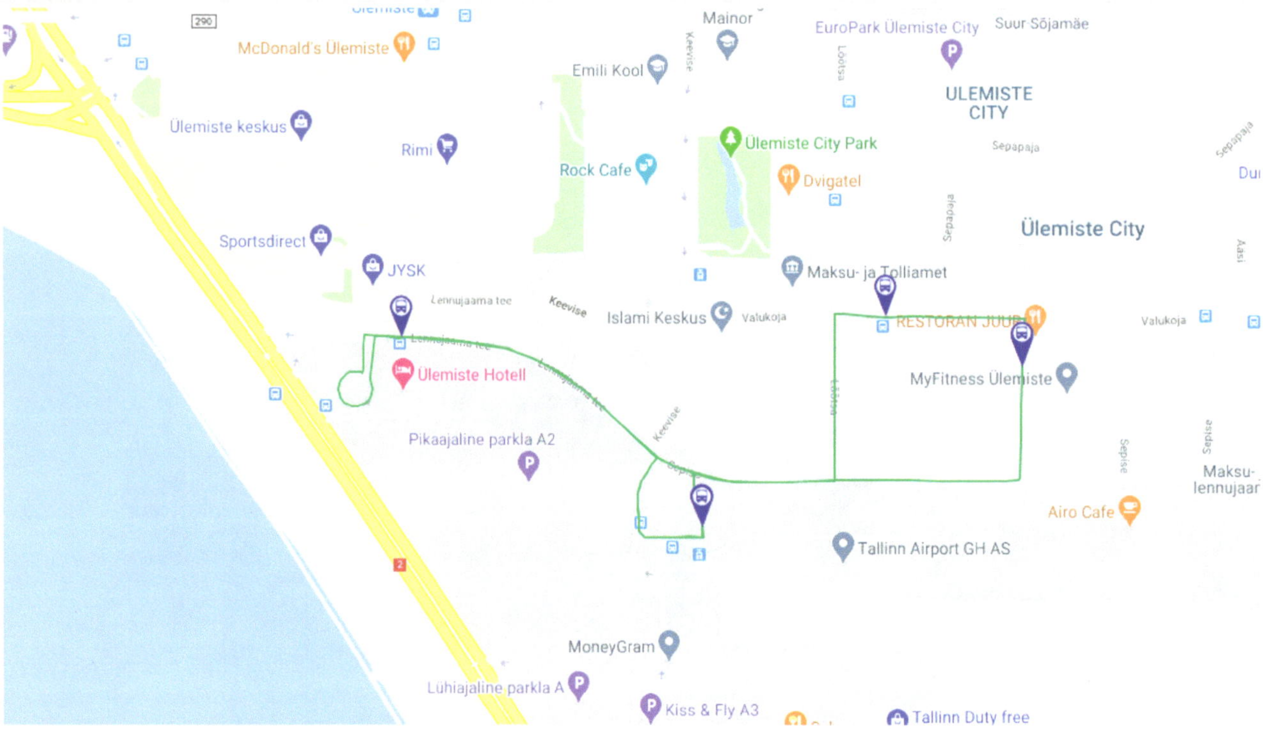

Reprinted with permission from Taltech

technology (IT). However, in the context of safety-critical applications, a methodology needs to be developed to manage safety as well as cybersecurity issues.

2. **Mass Customization**: One of the major advantages of open-source systems is the ability for mass customization driven by the energy of local teams. In the IT world, a simple example of this power is the language-specific translations of user interfaces. In the context of AV technology, it is likely that localized engagement will accelerate the exposure and solution development of a variety of climates as well as cultural situations.

Lake Nona, Florida

Large planned communities are quite popular in Florida with examples such as the Villages [38], On Top of the World [39], and Babcock Ranch [40]. These communities are large coherent ecosystems offering an overall "lifestyle" for the customer. Traditional services would include recreational facilities, interest clubs, and managed travel. Mobility is a critical factor for these communities with some offering a shadow transportation system for bicycles or golf carts. Recently, there has been a great deal of interest in these communities to extend the services model with automated mobility services. The Villages has a well-publicized relationship with Voyage [41] via a robo-taxi model, and Babcock Ranch [42] with a traditional shuttle-based approach. One of the most interesting and consistent experiments has been happening at Lake Nona, Florida (Figure 9).

Lake Nona is 17-square-mile planned community south of Orlando, Florida, whose development is led by Tavistock Development Corporation [43]. The Move Nona AV program launched in September 2019 with an inaugural route connecting residential neighborhoods to a shopping and dining district within the community. The initial route is 1.2 miles with three stops connecting Lake Nona Town Center Boxi Park, the Pixon Apartments, and The Laureate Park neighborhood village center (Figure 10). A private company, Beep, founded by IT veterans, drove the Move Nona program with funding from Tavistock, which had a desire to increase the differentiation and value of their offering to their customers (their primary monetization model continued to be real estate sales). Meanwhile, Beep pioneered layered revenue models within shuttle advertising.

The service currently runs seven days a week with extended hours during the weekends to accommodate passenger service demand. Multiple additional routes are being deployed this summer (2020), making Lake Nona one of the largest autonomous fleets in the world servicing a single community. As such, *it is one of the most robust AV solutions in the world*, and its success has led to a $20 million BUILD grant [43] from the US Department of

FIGURE 9. **Beep shuttle.**

Cari Rubin Photography/Shutterstock.com

FIGURE 10. **Lake Nona Beep route.**

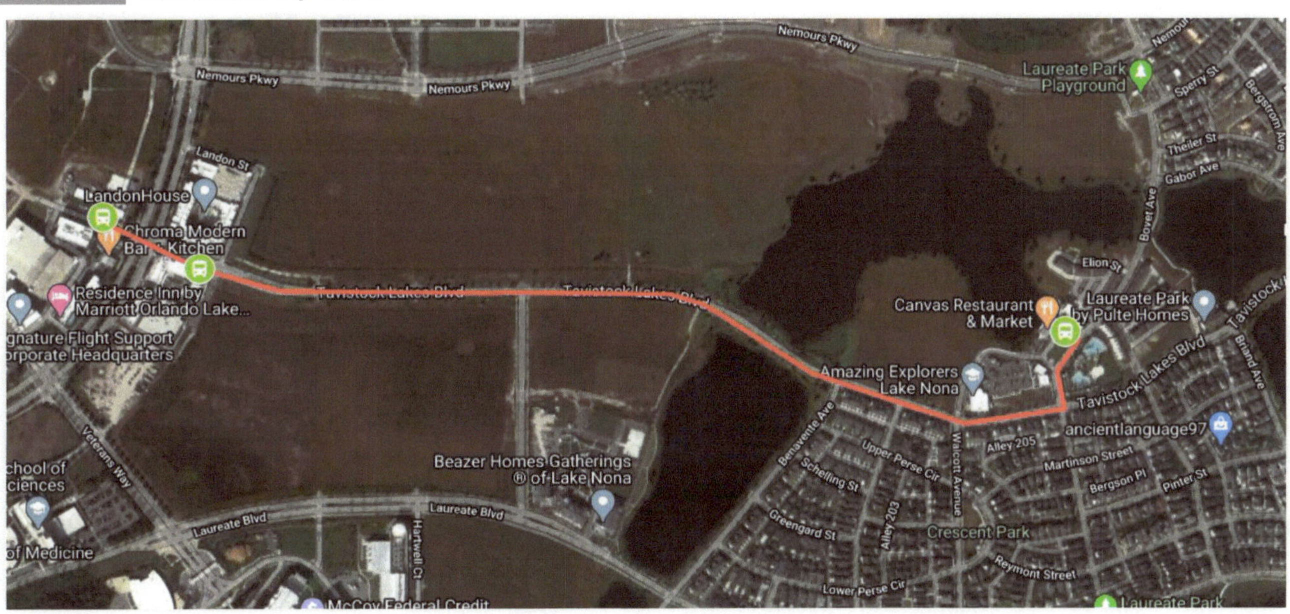

Reprinted with permission. Courtesy: Beep

Transportation. This grant will help expand the network and will also include a full-service mobility hub, linear park, linear park bridge, AV infrastructure, and bicycle transportation network culminating in the creation of more than 25 miles of multimodal path [44].

Similar to Jacksonville, Beep worked with regulators and vendors to qualify the solution. Given the nature of the technology, the most important lesson was that vendor relationships must be treated as partnerships if they are to be successful. The current state of the AV technology market does not support the rapid deployment of generalized solutions. Each platform is unique in its own way, with specific strengths and (oftentimes nonobvious) quirks, and each deployment or operational design domain (ODD) often provides its own challenges. Ensuring a deep understanding of each platform's limitations—as well as where the vendor may be able to address specific challenges—has been essential toward ensuring a safe, reliable, and high-quality experience for the passenger.

Recommendations

1. **Monetization model:** Lake Nona is an example of how a coordinated approach to implementing autonomous technologies as a component of a longer-term mobility master plan can potentially provide a near-term monetization model.
2. **AV Operators:** Perhaps the most interesting part of the Lake Nona experiment is the introduction of an intermediate AV operating vendor with Beep. This layer of capability seems necessary to fill the considerable gap between existing transportation agencies and AV vendors.

Summary

Public transportation shuttles show the promise of being a good application for AV technology because they operate at low speed, on fixed routes, and potentially gain dedicated resources. From a business perspective, there is already an existing public transit business model or there are opportunities for providing mobility services as a differentiator for a real estate development business model. Jacksonville, Singapore, Tallinn, and Lake Nona are actively engaging in employing this technology for their very unique circumstances.

As they engage, the reality of "rubber meeting road" exposes issues of the challenge of fitting into the complex machinery of an organic community, basic robustness issues with the vendor solutions, and gaps in knowledge between various communities. As these challenges are exposed, solutions are developed, and "growing pains" are overcome. Over the next few years, it is likely that public transportation will be one of the first places for robust delivery of autonomy technology.

SAE EDGE™ Research Reports

SAE EDGE™ Research Reports, like the present report on "Unsettled Topics Concerning Autonomous Public Transportation Systems," are intended to push further out into still unsettled areas of technology of interest to the mobility industry. SAE launches these reports before attempting to form a joint working group, let alone a cooperative research program or a standards committee.

SAE EDGE™ reports are intended to be quick, concise overviews of major unsettled areas where vital new technologies are emerging. An unsettled area is characterized more by confusion and controversy than established order. Early practitioners must confront an absence of agreement. Their challenge is often not to seize the high ground but to find common ground. These scouting reports from the frontiers of investigation are intended merely to begin the process of sorting through critical issues, contributing to a better understanding of key problems, and providing helpful suggestions about possible next steps and avenues of investigation.

SAE EDGE™ Research Reports, therefore, are fundamentally distinct from the more formal working groups approach and far removed from the more mature research program and standard's development process.

Next Steps for the AVs and the Transportation Ecosystem

This publication should be considered only as a first step toward clarifying the issues around AVs and the public transportation ecosystem. The intention behind this and other SAE EDGE™ Research Reports is to start a dialogue among interested parties on important industry-wide topics that require further attention. The expectation is that these explorations of unsettled areas of technology will lead to the formation of working groups and, ultimately, committees that can address and resolve the issues they raise, producing a framework for developing a common vocabulary of definitions, best practices, protocols, and standards needed to support continued progress toward safer and more innovative products.

The experts' collaboration that gave rise to this publication demonstrated a great willingness on the part of the industry to define the terminology, procedures, and eventually the standards needed to enable AV technology to move ahead as quickly and efficiently as possible.

This SAE EDGE™ Research Report on autonomous public transportation identifies the following key topics for further pursuit, both through continued informal discussions among industry practitioners and through more formal working groups:

- Validation and certification of autonomous public transportation systems

- Education regarding and certification of the public transportation ecosystem
- Cross-functional standards between public transportation and other public functions

Recommendations

The overall recommendations of this SAE EDGE™ Research Report can be summarized as follows:

1. **AV Education:** Public Transportation consists of many distinct entities and there is an intense need for education on the basics of autonomous technology.
2. **AV Engagement Model:** Autonomous systems are an emerging technology. Public transportation agencies must understand that a much more formal and staged process of introduction will be required as compared to more mature technologies. For large metropolitan areas, this may well involve dedicated test centers, which include specific features of the city and must adapt to changes in the city infrastructure/transportation/IoT conditions.
3. **Agency Teaming:** Public transportation sits in a complex ecosystem as a part of an overall city. Introduction of AV technology will require alignment with other city functions such as police, fire, and medical services.
4. **Open Source:** Given its power for crowd-sourced innovation, public transportation agencies should familiarize themselves with open-source methods. Open-source approaches offer the ability to customize locally but require access to technical expertise.
5. **Business Models:** For some time, autonomous technology will be more expensive than the conventional alternative. Public transportation agencies should look to enable such models to capture the value from AVs.
6. **Disaster Management:** AVs may take critical roles in transportation of people with no access to transportation in downtown or in regulation of the evacuation traffic, especially if they have their own dedicated lanes/routes.

Definitions

AI - Artificial Intelligence

AV - Automated Vehicle

BSIC - Bay Street Innovation Council

BUILD - Better Utilizing Investments to Leverage Development [Transportation Discretionary Grant]

IT - Information Technology

JTA - Jacksonville Transportation Authority

ODD - Operational Design Domain

OEM - Original Equipment Manufacturer

SSB - Sustainability Singapore Blueprint

U²C - Ultimate Urban Circulator

US - United States

UUC - Ultimate Urban Connector

Acknowledgments

Recognition should go first to all the participants, many of who also provided feedback on the draft version of this publication. Without their input and initiative, this SAE EDGE™ Research Report would not have been possible.

Ravio Sell, PhD, Tallinn University of Technology
Clayton Tino, PhD, Beep, Inc.
Bernard Schmidt, Jacksonville Transportation Authority
Mustafa Akbas, PhD, Embry-Riddle Aeronautical University
Jim Cherian, PhD, Centre of Excellence for Testing & Research of Autonomous vehicles—NTU
Niels de Boer, Centre of Excellence for Testing & Research of Autonomous vehicles—NTU

The authors of this document together with the SAE Team responsible for its creation join in expressing their deepest appreciation to all the individuals mentioned above.

Rahul Razdan, PhD
RazTech Research Institute

References

1. Iclodean, C., Cordos, N., and Varga, B.O., "Autonomous Shuttle Bus for Public Transportation: A Review," *Energies* 13(11):2917, 2020.
2. Azad, M., Hoseinzadeh, N., Brakewood, C., Cherry, C.R. et al., "Fully Autonomous Buses: A Literature Review and Future Research Directions," *Journal of Advanced Transportation* 2019:1-16, 2019.
3. Bucchiarone, A., Battisti, S., Marconi, A., Maldacea, R. et al., "Autonomous Shuttle-as-a-Service (ASaaS): Challenges, Opportunities, and Social Implications," arXiv preprint arXiv:2001.09763, 2020.
4. "About," Ultimate Urban Circulator (U²C)—Jacksonville Transportation Authority, u2c.jtafla.com/, accessed Oct. 16, 2020.
5. Bortzfield, B., "JTA Hires Automation Chief from Amazon; Tests Third Driverless Vehicle," *WJCT News*, news.wjct.org/post/jta-hires-automation-chief-amazon-tests-third-driverless-vehicle.

6. Hyatt, K., "Florida's New Bill Lets Driverless Cars Be Tested with Minimal Restrictions," Roadshow, CNET, June 16, 2019, www.cnet.com/roadshow/news/florida-self-driving-car-bill-hb311/, accessed Oct. 16, 2020.
7. "Our Home, Our Environment, Our Future," Sustainable Singapore Blueprint, www.mewr.gov.sg/ssb, accessed Oct. 16, 2020.
8. CSC, "Public Transport Planning and the Technological Revolution," www.csc.gov.sg/articles/public-transport-planning-and-the-technological-revolution, accessed Oct. 16, 2020.
9. Razdan, R., "Jacksonville's Journey in Engaging with Autonomous Vehicle Technology," *Forbes Magazine*, Mar. 6, 2020, www.forbes.com/sites/rahulrazdan/2020/01/25/jacksonvilles-journey-in-engaging-with-autonomous-vehicle-technology/.
10. "Centre of Excellence for Testing & Research of AVs—NTU (CETRAN)," Energy Research Institute @ NTU, erian.ntu.edu.sg/Programmes/IRP/FMSs/Pages/Centre-of-Excellence-for-Testing-Research-of-AVs-NTU-CETRAN.aspx.
11. Centre of Excellence for Testing & Research of Autonomous Vehicles—NTU (CETRAN), https://cetran.sg/, accessed Oct. 16, 2020.
12. Ministry-of-Transport, "Committee on Autonomous Road Transport for Singapore," www.mot.gov.sg/news-centre/news/Detail/Committee-on-Autonomous-Road-Transport-for-Singapore, accessed Oct. 16, 2020.
13. "NTU, LTA and JTC Unveil Singapore's First Autonomous Vehicle Test Centre," (n.d.), Retrieved Aug. 15, 2020, from https://www.jtc.gov.sg/news-and-publications/press-releases/Pages/20171122(PR1).aspx.
14. "Safety," *Ike*, www.ike.com/safety.
15. LTA, "New LTA.GOV.SG," www.lta.gov.sg/content/ltagov/en/newsroom/2019/10/1/Autonomous_vehicle_testbed_to_be_expanded.html.
16. Edb.gov.sg, www.edb.gov.sg/en/news-and-events/news/call-for-collaboration-on-pilot-deployment-of-autonomous-buses-and-shuttles.html, accessed Oct. 16, 2020.
17. "Singapore's First On-Demand Driverless Shuttle Buses to Ferry Passengers around Sentosa from Aug 26," todayonline.com, Aug. 20, 2019, https://www.todayonline.com/singapore/singapores-first-demand-driverless-shuttle-buses-ferry-passengers-around-sentosa-aug-26.
18. ST Engineering, "First On-Demand Autonomous Shuttle Trial Opens to Public," Aug. 20, 2019, https://www.stengg.com/en/newsroom/news-releases/first-on-demand-autonomous-shuttle-trial-opens-to-public/, accessed Oct. 16, 2020.
19. "ComfortDelGro Self-Driving Shuttle Bus Showcases New Features at Transport Congress," todayonline.com, Oct. 22, 2019, https://www.todayonline.com/singapore/demand-self-driving-shuttles-comfortdelgro-showcase-new-features-international-transport.
20. NUSMART Shuttle, "NUSmart Shuttle (EasyMile EZ10) Begins Passenger Service Trial at NUS," https://uci.nus.edu.sg/oca/latest-news/nusmart-shuttle-a-fully-autonomous-vehicle/, accessed Oct. 16, 2020.
21. Razdan, R., "Linux Beat IBM, Will Open-Source Software Beat Waymo and Tesla?" *Forbes Magazine*, Apr. 7, 2020, www.forbes.com/sites/rahulrazdan/2020/04/04/linux-beat-ibm-will-open-source-software-beat-waymo-and-tesla/.
22. "TalTech," Avaleht, www.taltech.ee/.
23. "Iseauto," YouTube, www.youtube.com/channel/UCBNkOeSC1HmDJ7PyHRP8fiA.
24. "TalTech Demonstrates Technological Excellence at Robotex," Nov. 28, 2018, https://old.taltech.ee/taltech-demonstrates-technological-excellence-at-robotex.
25. "Welcome!," Silberauto, www.silberauto.com/.
26. Salumae, T., "ISEAUTO Projekt (Iseauto.ttu.ee)," Iseauto.ttu.ee, 2017, iseauto.ttu.ee/wp-content/uploads/2017/08/tarkvararaamistik_iseauto.ttu_.ee_.pdf, accessed Oct. 16, 2020.
27. "Self-Driving Cars Technology & Solutions from NVIDIA Automotive," *NVIDIA*, www.nvidia.com/en-us/self-driving-cars/.
28. Comma.ai, "Introducing Openpilot," comma.ai/.
29. Apollo, apollo.auto/, accessed Oct. 16, 2020.
30. Autoware.AI, www.autoware.ai/, accessed Oct. 16, 2020.
31. Razdan, Rahul, "Will a Small Open-Source Effort from Japan Disrupt The Autonomous Space?" *Forbes Magazine*, Apr. 5, 2020, www.forbes.com/sites/rahulrazdan/2020/04/04/will-a-small-open-source-effort-from-japan-disrupt-the--autonomous-space-/.
32. "The First Ride of TalTech's Self-Driving Car Was a Success," Oct. 16, 2018, https://old.taltech.ee/the-first-ride-of-taltechs-self-driving-car-was-a-success.
33. Sell, R., Leier, M., Rassõlkin, A., and Ernits, J.-P., "Autonomous Last Mile Shuttle ISEAUTO for Education and Research," *International Journal of Artificial Intelligence and Machine Learning (IJAIML)* 10(1):18-30, 2020.
34. "ISEAUTO Operation in Snow," Video, https://photos.app.goo.gl/M6GXBC36ekUf5qDy5.
35. Auve Tech, auve.tech/, accessed Oct. 16, 2020.
36. Autonomous Vehicles Lab, autolab.taltech.ee/, accessed Oct. 16, 2020.

37. Seltz-Axmacher, S., "The End of Starsky Robotics," Medium, Starsky Robotics 10-4 Labs, Mar. 19, 2020, medium.com/starsky-robotics-blog/the-end-of-starsky-robotics-acb8a6a8a5f5, accessed Oct. 16, 2020.
38. TheVillages.com, (n.d.), Retrieved Aug. 15, 2020, from https://www.thevillages.com/.
39. "On Top of the World Is the Premier Retirement Community in Florida," Aug. 3, 2020, Retrieved Aug. 15, 2020, from https://ontopoftheworldcommunities.com/
40. "Sustainable City: Babcock Ranch, Florida," (n.d.), Retrieved Aug. 15, 2020, from https://www.babcockranch.com/.
41. Korosec, K., "Voyage Raises $31 Million to Bring Driverless Taxis to Communities," Sept. 12, 2019, Retrieved Aug. 15, 2020, from https://techcrunch.com/2019/09/12/voyage-raises-31-million-to-bring-driverless-taxis-to-communities/.
42. (n.d.), Retrieved Aug. 16, 2020, from https://www.builderonline.com/building/autonomous-vehicles-come-to-babcock-ranch_o.
43. Lake Nona Town Center, (n.d.), Retrieved Aug. 15, 2020, from https://tavistockdevelopment.com/project/lake-nona-town-center/.
44. Move Nona, (n.d.), Retrieved Aug. 15, 2020, from https://www.go-beep.com/move-nona-project/.

Contact Information

EDGEresearch@sae.org.

www.ingramcontent.com/pod-product-compliance
Lightning Source LLC
LaVergne TN
LVHW070941210926
848482LV00003B/279